D0849376

Nature's Children

ORCAS

Geoff Miller

GROLIER
EDUCATIONAL

FACTS IN BRIEF

Classification of Orcas

Class:	*Mammalia* (mammals)
Order:	*Cetacea* (whales)
Family:	*Delphinidae* (dolphin family)
Genus:	*Orcinus*
Species:	*Orcinus orca*

World distribution.   All the world's major seas and oceans.

Habitat.   Open ocean or near to coast; largest populations are found in cooler waters.

Distinctive physical characteristics.   Streamlined bodies; large dorsal fin; mostly black with white markings on the underside and along the flanks; large horizontal tail flukes and rounded flippers; blowhole on top of the head.

Habits.   Live in large family groups of males, females, and young called pods. Pods hunt together as a pack. Communicate using clicks, squeaks, and whistles. Playful and intelligent.

Diet.   Fish, squids, seals, dolphins, and penguins.

© 1999 Brown Partworks Limited
Printed and bound in U.S.A.
Editor: James Kinchen
Designer: Tim Brown

Published by:

GROLIER
EDUCATIONAL

**Sherman Turnpike, Danbury,
Connecticut 06816**

Library of Congress Cataloging-in-Publishing Data
Orcas.
    p. cm. -- (Nature's children. Set 6)
  ISBN 0-7172-9364-5 (alk. paper) -- ISBN 0-7172-9351-3 (set)
  1. Killer Whales--Juvenile Literature. [1. Killer Whales.] I. Grolier Educational (Firm) II. Series.
QL737.C432M55 1999
599.53'6—dc21
                                                                    98-33398

Contents

Orca Territory   Page   6

Giants of the Sea   Page   9

Whale Watching   Page 10

One of the Mammals   Page 13

Mighty Breaths   Page 14

Champion Swimmers   Page 17

Keeping Warm   Page 18

Unusual Noses   Page 21

Whale Blows   Page 22

Living Together   Page 25

Feeding Time   Page 26

Team Players   Page 29

Clever Creatures   Page 30

Orca Talk   Page 31

Finding a Mate   Page 34

A New Arrival   Page 37

Growing Up   Page 38

Fun and Games   Page 41

A Well-Deserved Rest   Page 42

Beached Whales   Page 45

Long Lives   Page 46

Words to Know   Page 47

Index   Page 48

When you go on vacation to the beach, do you ever stop to think about all the creatures living in the ocean? If you have been for a paddle in the sea, you may have seen small fish swimming around your feet. Perhaps you have found little crabs in rockpools at low tide. The sea is also home to far larger creatures, like the huge whales.

Perhaps you have seen a picture of a whale or seen a live whale close up in a marine theme park. One type of whale, the blue whale, is the largest animal in the world. Orcas, or killer whales, are part of a group of small whales called the dolphin family. Most of the members of this family of animals, such as the bottlenose dolphin, do not get called whales at all. Killer whales, however, are the largest members of the group and deserve their title. Let's find out more about these remarkable creatures.

Cool coastal waters are a favorite place for orcas to swim and hunt.

Orca Territory

Orcas live in oceans all over the world, including the great Atlantic, Pacific, and Indian Oceans. They can also be found in the cold, icy waters of the Arctic, in the far north, and the Antarctic, in the far south. Orcas prefer the shallow coastal waters around our continents, but they are also found many hundreds of miles from land, in very deep waters. Although they are sometimes seen in the warmer seas near the Equator, they live in greater numbers in cooler waters.

Around the continent of North America you can see orcas off the northwest coast near Puget Sound, in Washington State, and around Vancouver Island. You can also see them off the northeast coast, from Maine to Newfoundland. Orcas also live in the seas off Alaska, Hawaii, California, and Mexico.

An orca surfaces beside floating ice in the Antarctic Ocean.

Giants of the Sea

Orcas are not the largest of whales, but they can grow to a considerable size. A male killer whale can grow to 33 feet (10 meters) long and weigh as much as 13 tons (10,000 kilograms). That's as long as a small truck and the same weight as about 120 people! Fully grown male orcas are usually about a third bigger than females.

One of the orca's most striking features is its tall, triangular-shaped fin, which is about halfway down its back. Called the dorsal fin, it can be six feet (two meters) high in males, which is taller than an adult person. A large part of a whale's weight is made up of the thick layer of fat, or blubber, which keeps its body warm in the water.

This orca is swimming alongside a smaller member of the whale group—a dolphin.

Whale Watching

If you went on a whale-watching expedition, would you be able to recognize an orca? In fact, killer whales are quite easy to spot because of their distinctive black-and-white markings. Mostly they are black, but their bellies are white. They also have a finger-shaped white marking on their sides and a white, oval-shaped spot just above and behind each eye. The dull gray patch over an orca's back is less noticeable. These patterns are similar on all orcas, although occasionally you might see an albino orca, which is completely white.

Experts who study killer whales in the wild can recognize individual whales because every one has slightly different markings. When an individual whale has been identified, its movements can be studied. This helps us to learn more about how they behave.

An orca's fin stands high above the surface of the sea—a magnificent sight for these whale watchers.

One of the Mammals

Is an orca a type of fish? You could easily make the mistake of thinking that killer whales are a kind of fish. In fact, like all whales, orcas are mammals. Mammals are those animals that breathe air, give birth to live young, and feed, or suckle, their babies on the mother's milk. Can you think of any other mammals?

Most mammals live on land, such as monkeys, tigers, elephants, and humans. Orcas are known as marine, or aquatic, mammals. They spend their lives at sea, occasionally coming to the surface to breathe. Fish breathe using gills, which take in oxygen from the water, but whales are like us—they have lungs for breathing air. Another difference between whales and fish is in the shapes of their tails. The flat part of a whale's tail, called a fluke, is horizontal, or level with the water's surface, whereas the tail of a fish is vertical, or stretches up and down.

Opposite page:
A playful orca launches itself above the surface to breathe air.

13

Mighty Breaths

If you want to swim under water, first you have to take a very deep breath and then hold it. Holding your breath for any length of time is difficult—perhaps you can manage it for 30 seconds or even for as long as a minute? Orcas often need to take long dives in search of food, so they need to hold their breath for much longer than we can manage.

Orcas can hold their breath for up to 20 minutes at a time. You might think that they can do this because they have very big lungs, but compared to the size of their huge bodies, their lungs are really quite small. However, they do use their lungs very efficiently, breathing in lots of fresh air at the surface before taking a long, deep dive.

Orcas can stay under water for up to 20 minutes at a time.

Champion Swimmers

Can you swim? Swimming does not come naturally to humans, so learning to swim can be difficult. Some people practice hard and go on to win Olympic medals for swimming, but it is not easy. However, if we were more like orcas, we could probably win every swimming prize going!

As you might expect, orcas are very good swimmers. They are well adapted to living in the water. Their sleek and streamlined bodies glide easily and quickly through the sea at speeds of up to 24 miles per hour (38 kilometers per hour). The tall dorsal fin acts as a kind of stabilizer, like the keel of a boat, helping orcas to keep their balance. Rounded flippers shaped like paddles are ideal for steering. To propel their heavy bodies through the water, orcas' strong muscles move their powerful tail flukes up and down.

Showing its power and agility, a large male orca leaps out of the water with ease.

Keeping Warm

Animals that live on land usually have hairy bodies or thick fur to keep them warm, while birds have feathers. So how do orcas manage to stay warm in the cold oceans in which they live? Like us, whales are warm-blooded creatures and have a body temperature of about 95°F (35°C), similar to our own. We cannot survive for very long in the sea because our bodies lose heat much faster in the water than on dry land, and we quickly feel the cold. Whales have a special layer of fat, called blubber, to get around this problem. This thick layer lies under the orca's delicate outer skin. The fat prevents heat loss and stops the whale from getting cold. As much as a third of a whale's total body weight consists of blubber.

A thick layer of blubber under the skin helps orcas keep warm—even in the ice-cold waters of Antarctica.

Unusual Noses

Orcas and other whales breathe in the air they need through a blowhole on the top of their head. The blowhole is a whale's nose, or nostril. You may think it is a strange place for a nose, but for a whale it makes perfect sense. It means that when an orca surfaces after one of its long dives, its blowhole is the first part of its body to emerge from the water. Also, it allows the orca to swim and breathe at the same time, which is very useful as it spends 95 percent of its time under water!

When orcas dive beneath the water, special lips close tightly over the blowhole to stop any water from getting in. Only when the orca is ready to take a breath are the lips drawn back to let the whale breathe again.

Orcas breathe through a single nostril, called a blowhole, which is on top of their head.

Whale Blows

Have you ever seen a drawing or a painting of a whale? Such pictures often show a fountain of water shooting up out of the whale's head. However, orcas and other whales do not really spout water like this, although the way they breathe out can make it look as if they do. Following a long dive, they need to blow out all the used air from their lungs. Thousands of tiny drops of moisture in the whale's warm breath quickly cool down in the air, making a big cloud of vapor. This is the well-known "blow" that a whale makes when it surfaces.

Expert whale watchers are able to tell one type of whale from another by the different shapes of blow they make. Sometimes the blow is accompanied by a loud noise as the air blasts out and upward.

With a cloud of water vapor and a loud blast of air the orca "blows" a breath out.

Living Together

Orcas are friendly animals and like to live together in family groups. A family of orcas is called a pod. A pod is normally made up of between five and 20 individual whales, although some pods can have as many as 55 members. Sometimes several pods get together to form an even larger group of whales known as a herd.

An orca pod usually consists of male and female adults, young orcas, called juveniles, and baby orcas, or calves. A family group stays together for many years and may even remain together for life. There are lots of advantages in living together in a pod. All the family members can help each other when hunting for food. Grown-up female whales can help an orca mother to look after her calf until it is old enough to look after itself.

Members of an orca family stay close together as they travel.

Feeding Time

Killer whales are meat eaters, and as their name suggests, they kill other animals to eat their flesh. Orcas eat most kinds of fish, squid, seals, sea lions, porpoises, dolphins, sea birds, and penguins. Occasionally they attack and eat other kinds of whale, even ones much larger than themselves. Although killer whales have a fearsome reputation, there is no record of an orca killing or even attacking a human being.

Orcas have huge appetites—one killer whale is said to have had the remains of 13 porpoises and 14 seals inside its tummy! Orcas do not chew their food. They use their 40 to 50 sharp teeth to rip apart the larger animals they catch before swallowing big bites. Their tummies do the rest of the work, digesting these pieces of meat.

Keiko, the orca who starred in the movie Free Willy, *shows off his mouthful of teeth.*

This orca is flapping its tail to round up a shoal of herrings, ready to eat for dinner.

Team Players

Orcas are big eaters and always on the prowl for food. An orca pod will cover great distances in its search for food, sometimes traveling 60 miles (100 kilometers) in a day. They hunt together, often cleverly planning their actions before moving in for the kill.

With the exception of the youngest whales, all the orcas in the group take part. Using sounds to communicate with each other, the orcas seem to use a deliberate plan—first tormenting and then attacking their prey. Many types of fish swim together in large groups, or shoals. Killer whales have often been seen rounding up a shoal of fish, sometimes trapping it between the seashore and the pod members. Once they have been gathered together, the fish are eaten one by one until the orcas' tummies are full.

Clever Creatures

Although killer whales are mainly thought of as fierce hunters, they are also extremely intelligent animals. Like dolphins and porpoises, orcas have very large brains. Their intelligence may be more like ours than any other animal's. Scientists think that whales have developed their own system of communication, too, and may be able to think, learn, and remember information.

Orcas have good eyesight, but their hearing is even better. Because there is not much light under the water, it is difficult for orcas to communicate with, find, or follow one another by sight. So orcas have developed a range of different sounds to communicate with each other. By making noises and listening for the sounds of other whales in the pod, they can find their way around and stay in touch in the murky waters.

Orca Talk

Killer whales make a variety of different noises: chirps, high-pitched whistles, buzzes, and clicks. Some orca sounds cannot be heard by the human ear, although scientists have detected them using special equipment. Orcas make their noises by forcing air through a tube in their heads, close to the blowhole.

Orcas use sounds to locate their prey in a special way, called echolocation. They emit a stream of sounds that travels through the water. The sounds are reflected back off any nearby fish and return to the orca as echoes. The echoes are like those you hear when you shout at a wall or building—your voice comes bouncing back a fraction of a second later.

Each orca pod may have its own series of sounds. Whales within a pod can understand the sounds of other pod members, but calls from another family sound slightly different.

Finding a Mate

When they are about eight or nine years old, female orcas are grown up enough to have babies. Depending on where they live, killer whales have babies at different times of the year. In the North Atlantic Ocean they mate mostly between October and November, while in the northern Pacific Ocean the popular time for finding a mate is in the spring and summer months.

Before they mate, male and female orcas will often frolic with each other in a lively display of touching and teasing. Sometimes more than one male orca will chase a single female and try to mate with her. Once she is pregnant, the female will have to carry her baby for about 15 months, which is six months longer than your mom carried you.

A female orca with her shorter, curved fin and a male with his taller, triangular fin travel side by side.

A New Arrival

The orca mother will almost always have only one baby. The newborn whale is called a calf. Orca calves are born into their watery world tail-first. If they were born head-first their natural instinct to breathe straight away might drown them.

Infant orcas are exactly like their parents, just miniature versions. They are born with their eyes wide open and their senses alert. Their first priority is to get to the surface of the water to take their first breath of air. Sometimes a mother helps her calf by giving it a gentle nudge up to the surface. Once the first breath has been taken, the orca mother gives her baby its first feed of milk from her teats. Then, following its instincts, the calf starts to swim beside its mother, never straying far from her side.

As soon as an orca is born,
swimming skills come naturally—an
orca calf does not have to learn.

Growing Up

At birth orca calves are about seven feet (two meters) long and weigh around 400 pounds (180 kilograms). That's longer than an adult person is tall, and about the same weight as two grown-up humans.

The orca mother continues to suckle the youngster, nursing her calf for about a year. Orca pods are very protective toward their new members. When calves and young orcas leave the main group to play, an adult killer whale usually stays near to the youngsters, keeping a watchful eye on them. Occasionally a calf's mother will swim off by herself, leaving her baby with an adult male orca. Often the calf will carry on playing, and the male adult joins in the fun. Later the female returns to the pod to look after her calf again.

An orca calf swims in safety between two grown-ups.

Fun and Games

Killer whales are very boisterous and playful creatures. Captive orcas in marine parks are easily taught to do tricks by their trainers. But they do not need any encouragement from humans to play around! In their natural ocean environment they are often seen performing acrobatics for fun.

Orcas of all ages like to play, but the younger ones in particular seem to enjoy frolicking in the water. They leap, jump, and slap the water with their tail flukes. Sometimes they "spyhop"—rising head-first out of the water as far as their flippers, so that both their eyes are above the waterline. Orcas' curious behavior at the surface of the sea is fascinating to watch. We can only guess at what they are up to during their long, deep dives below.

Like an acrobat in the open ocean,
a wild orca twists and turns,
slapping the water with its tail.

A Well-Deserved Rest

Do you get tired after playing for a long time? Orcas do, so after a busy time spent playing and hunting, they like to take a rest. Like any other animals, orcas need to sleep. Studies of orcas in the wild show that they sleep for a minimum of half an hour and a maximum of around six hours each day. Usually they sleep for about two hours at a time. While sleeping, they continue to swim, but much more slowly than usual.

Orcas like to do things in a group; they even rest and sleep together. They swim slowly, staying only a few feet apart—some of them even stay within touching distance. They control their breathing so that they can rise to the surface within a short time of each other. Every three or four minutes they all emerge, taking several breaths before submerging again to continue their quiet doze.

This orca is taking a well-earned rest, dozing just below the surface of the water.

Beached Whales

What kind of dangers does a killer whale face? A real danger for orcas and other whales is "stranding." In many parts of the world whales often become stranded, or washed up, on beaches. If they cannot get back to the sea, they are likely to suffer from overheating and starvation, and they usually die.

Whale strandings are a mystery. No one knows for sure what causes these sad events. Some scientists think that it happens if a whale loses its way. Whales' sense of direction is usually very good, but it may sometimes fail, causing the whales to lose their way. If they get lost, the result can be a disaster. Because orcas live in groups, they will almost always follow the leading animal. If the leader is lost, the whole pod can become stranded.

Many orcas swim close to the coast without getting stranded. Why some do is a mystery.

Long Lives

How long do you think a killer whale can live? As they have no natural enemies and no other animals hunt them for food, orcas usually enjoy long lives. We do not know exactly how long they live, but experts think that female killer whales live to around 50 years of age, while males probably live to about 30. It is possible that some orcas live even longer—for 50 or 60 years.

Although orcas are not at risk from other hunting animals, they do face several dangers. People have hunted and killed whales for hundreds of years, mainly for the oils that can be extracted from their blubber. Another danger lies in being accidentally caught in fishing nets, where they can die from lack of air. The careless dumping of poisonous wastes into the ocean can also injure a whale. Although many orcas survive today, we must take care of their ocean world so that these special animals will thrive in the future, too.

Words to Know

Albino   Animal without natural coloring.

Calf   Baby orca.

Dorsal fin   Fin on the animal's back.

Echo   A sound heard after being reflected off a solid object.

Equator   An imaginary circle around the earth, midway between the North and South Poles.

Fluke   The wide, flat part of a whale or dolphin's tail.

Gills   The organs that fish use to breathe.

Keel   Stabilizing fin under a boat.

Mammal   Warm-blooded animal that suckles its young.

Prey   Animal that other animals hunt for food.

Shoal   Large group of fish swimming together.

Spyhop   To rest in an upright position, partly out of the water. Orcas spyhop to look around above the surface of the ocean.

Strand   To run aground. Many whales strand on beaches, but no one is sure why.

Streamlined   Smooth and sleek, able to move easily through air or water.

Vapor   Cloud of mist or other gas.

INDEX

baby orca, *see* calf
birth, 37
"blow," 22; *illus.*, 23
blowhole, 21, 31; *illus.*, 20
blubber, 9, 18
body temperature, 9, 18
breathing, 13, 14, 21, 37
breeding, 13, 34

calf, 25, 34, 37, 47; *illus.*, 39
color, 10
communication, 29, 30, 31

diet, 26
distances traveled, 29
distribution, 6
diving, 13, 21, 22
dorsal fin, 9, 17, 47; *illus.*, 11, 35

echolocation, 31, 47
eyesight, 30

feeding, 14, 26, 28, 37, 38
flippers, 17
fluke, *see* tail

hearing, 30
herd, 25
human threat, 46
hunting, 29; *illus.*, 28

intelligence, 30

length, 9, 38
life span, 46

mammal, 13, 47
mating, 34

noise, 22
nose, 21

playfulness, 41
pods, 25, 29, 31, 38, 45
prey, 26, 29, 31, 47

shape, 17
size, 9
sleep, 42; *illus.*, 43
sounds, 30, 31
speed, 17
spyhop, 41, 47
stranding, 45
swimming, 17

tail, 13
teeth, 26; *illus.*, 27

weight, 9, 38

young orcas, 25

Cover Photo: Brandon D. Cole / Corbis
Photo Credits: Andy Rouse / NHPA, page 4; Wolfgang Kaehler / Corbis, page 7; Michael Melford / Imagebank, pages 8, 27; T. Kitchin & V. Hurst / NHPA, pages 11, 12, 40, 44; Amos Nachoum / Corbis, pages 15, 28; Brandon D. Cole / Corbis, pages 16, 23, 36; A.N.T. Photo Library / NHPA, page 19; Galen Rowell / Corbis, page 20; Natalie Fobes / Corbis, page 24; Gerard Lacz / NHPA, page 32; Stuart Westmorland / Corbis, page 35; David E. Myers / NHPA, page 39; Rob Atkins / Imagebank, page 43.